W9-DIW-334

Practical Plants

written by
Joyce Pope

Facts On File
New York • Oxford • Sydney

Published in the United States in 1990
by Facts On File, Inc., 460 Park Avenue South,
New York, NY 10016

A Templar book
Devised and produced by The Templar Company plc,
Pippbrook Mill, London Road, Dorking,
Surrey RH4 1JE, Great Britain

For information contact: Facts On File, Inc.,
460 Park Avenue South, New York, NY 10016

Library of Congress Cataloging-in-Publication Data
Pope, Joyce.
    Practical Plants / Joyce Pope.
        p.   cm. -- (Plant life)
    Includes bibliographical references.
    Summary: Surveys the many uses of plants, including
nutrition, fuel, building, gardening, and medicine.
    ISBN 0-8160-2424-3
    1. Botany, Economic--Juvenile literature.   2. Plants,
Useful--Juvenile literature.   [1. Plants, Useful.]   I. Title.
II. Series.
SB107.P65   1990
581.6'1--dc20   90-32394
                                                             CIP
                                                             AC

Facts On File books are available at special discounts
when purchased in bulk quantities for businesses,
associations, institutions or sales promotions. Please call
our Special Sales Department in New York at
212/683–2244 (dial 800/322–8755 except in NY, AK or HI).

Notes to Readers
There are some words in this book that are printed in
bold type. An explanation of them is given in the
glossary on page 58.

Editor Wendy Madgwick
Designer Mike Jolley
Illustrator Avril Turner

Color separations by Positive Colour Ltd, Maldon, Essex
Printed and bound by L.E.G.O., Vicenza, Italy

10 9 8 7 6 5 4 3 2 1

Contents

Uses of Plants

All plants are practical plants. Each one is a mini-factory that uses the power of the sun to change nonliving chemicals into its own living tissues. Any that did not do this efficiently would become extinct.

As the world's basic converters of energy, plants have practical uses for animals. The first of these is as food. Without the food that plants produce no animal could live. Even the flesh-eaters depend on plant feeders as their prey. Plants also produce the gas oxygen as a waste product of their activity. Animals must have this gas to breathe. So there are two main ways in which plants are of practical use to animals.

People and plants

Human beings, like all other animals, need to breathe and to eat. But people also use plants for many other things. We use their fibers to make fabrics, their **pigments** to make dyes and their chemicals to make drugs. We make and furnish our homes from parts of plants and we use them to beautify ourselves and our lives. All of the major groups of plants are used to some

extent but by far the most important are the flowering plants. Many thousands of different species are used throughout the world for a host of purposes. This book is about the many ways that people make use of plants.

Nonflowering plants

Some algae, which include the seaweeds, are eaten, and the **algenates** that they produce are used as stiffeners in many products from pudding mixes to carpet backing. **Lichens** are used to make dyes, and in some parts of the world some species are valued because they are good to eat. Some **mosses** are used for packing, and many club mosses are used for their medicinal properties. **Horsetails** and **ferns** are eaten in some parts of the world and have also been used to treat many kinds of illness. The rough stems of some species of horsetails make a good substitute for sandpaper.

Flowering plants

We can use almost every part of flowering plants. In some, such as carrots or turnips, we use the roots. The stems may be valuable as food in plants such as celery, while in woody plants, the trunk of a tree is the most important source of timber. The bark may be used for fiber and the sap for sugars or gums. We eat the leaves of many plants, such as cabbages, and we feed the leaves of others to livestock. The spice clove is a dried flower bud, and apart from growing plants for the beauty of their flowers, we use them for their fragrance. We eat fruits of many kinds and use others for obtaining oil or chemicals. Seeds are also an important food source, but they have many other uses from making medicines to making necklaces.

◀ *We grow flowers in our gardens because they are lovely to look at and smell pleasant.*

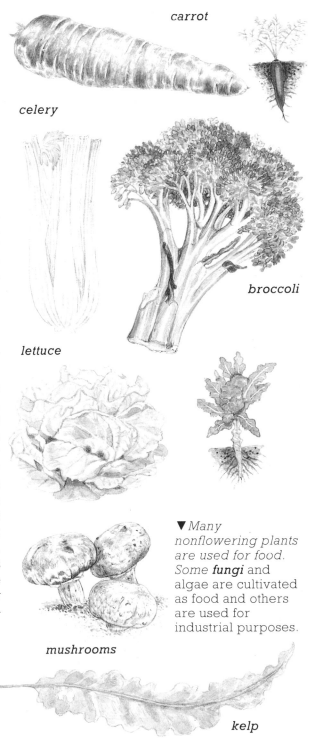

▼ *We can eat the stems, leaves, roots and flowers of many food plants.*

carrot

celery

broccoli

lettuce

▼ *Many nonflowering plants are used for food. Some **fungi** and algae are cultivated as food and others are used for industrial purposes.*

mushrooms

kelp

The Importance of Grass

To many people grass is the stuff that lawns are made of, and beyond that it is a weed. But the grass family includes plants that are more important to us than any other. All of the world's major civilizations of the past and present have been founded on grass, for the cereals that feed us are cultivated grasses.

Farmers began to grow cereals in various parts of the world between about 10,000 and 5,000 years ago. This is called the **Agricultural Revolution**, and it changed the lives of human beings more than any other single thing in history. Until that time most people spent the greater part of their lives hunting or looking for food. After the Agricultural Revolution some people became farmers, while others lived in villages and towns and became traders and manufacturers. We still have that pattern of life today.

Wheat

No other sorts of plant give us so much food or are put to such a variety of uses as the grasses. The most important is wheat. It is now grown throughout much of the world outside the tropics, but it probably came originally from a dry area somewhere east of the Mediterranean. There are now thousands of varieties of wheat, which vary so much that nobody knows exactly what their wild ancestor was like. There seem to be three main groups, each

▼ Emmer wheat was widely grown in ancient times and has been found preserved in Egyptian tombs.

▼ Durum wheat is grown in the Mediterranean region to produce flour for macaroni, spaghetti and other pasta.

▼ Einkorn wheat was probably one of the ancestors of cultivated wheat. It has fairly small grains and is difficult to thresh.

▼ Hard wheat makes the best bread. Soft-grained wheat is used to make flour for cakes and pastry.

▶ Scientists have bred many varieties of wheat in order to produce heavier crops with more disease-resistance.

of which is different genetically, so it is possible that modern wheats are **hybrids** and **mutants** entirely different from any wild plant.

All-purpose food

The most important kind of wheat is bread wheat. This is the common wheat grown in the USA, Australia, the USSR and Europe. The best bread wheat is hard and has a good deal of **gluten**, which helps to make bread dough sticky and elastic. Areas with cool, wet summers produce a softer grained version, which is not so good for making bread but can be used to make all-purpose flour and breakfast cereals. Another variety produced in warmer, drier places is durum or macaroni wheat. It is very hard, and has a very high gluten content. As its name suggests, it is used for making pasta. It is a "bearded" wheat, for as it grows each grain has a long stiff hair or **awn** protecting it.

FOOD FOR THOUGHT

*C*ereal **grains** contain the food for the next generation of grasses. This includes **starch**, a little **protein**, **minerals** and **vitamins**. We use grains whole, or grind them into flour. They are also used to feed livestock such as chickens and cattle. Without cereal grains you could not have an egg for breakfast or a milkshake at lunchtime or a steak for dinner.

Barley, Oats and Rye

Barley is still found as a wild plant in parts of Turkey and Syria. It is much hardier than wheat and is often grown on poorer soils. It was first cultivated in the Near East and around the Mediterranean about 10,000 years ago.

Today barley is less important because it contains very little gluten and cannot be used for making light, well-risen bread, although it is still eaten where unleavened bread is popular. About half of the crop is fed to livestock and a little goes to make pearl barley or barley meal. The other important use is for malting to make beers and whiskys. A by-product of this is yeast, which is used for baking.

Oats

Oats were probably found as weeds among the other crops before they were grown for their own value. There are many varieties, some of which can thrive in poor soils and climates.

Oats are not used for making bread, but oatcakes and porridge are eaten in some areas. Oat oil is used in some other breakfast foods. Oats provide high-energy food for horses. Before farming became mechanized, a large percentage of farmland was cultivated to grow oats for fodder. Oats are still used for feeding animals, either directly or as pellets.

An unexpected use is made of oat husks. At one time these were burned or made into packing materials. Now a solvent, **funfural**, is made from them. This is used by the oil industry and in many modern products, including man-made fibers, synthetic rubber and antiseptics.

Rye

Rye was probably a weed in early crops of wheat and barley. As these warmth-loving plants were cultivated further and further

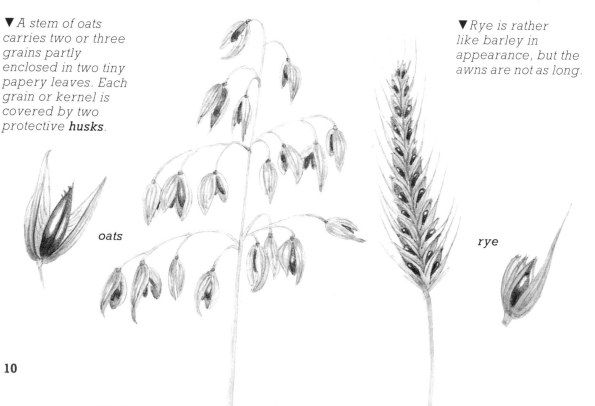

▼ *A stem of oats carries two or three grains partly enclosed in two tiny papery leaves. Each grain or kernel is covered by two protective* **husks**.

oats

▼ *Rye is rather like barley in appearance, but the awns are not as long.*

rye

north, the rye, which is hardier, formed a larger proportion of the crop. Eventually it was grown as a crop, especially in climates too harsh to grow other cereals. Today rye is mostly grown in eastern Europe and parts of North America. The grain contains about 13 percent protein and is suitable for making bread, though wheat flour is sometimes mixed with it.

Another use of rye is in the making of vodka and rye whisky. The grain and sometimes the plant are used as animal feed. The straw is used for bedding animals, thatching and making paper.

ST. ANTHONY'S FIRE

*When hardier varieties of wheat and barley were produced, rye lost some of its popularity as a crop. One reason for this may be that it is sometimes attacked by a **fungus**, causing a disease named **ergot**. In the Middle Ages it was known as St. Anthony's fire. It causes hallucinations and often death in humans and cattle. In some areas today rye is grown just for the ergot, which is used by the pharmaceutical industry.*

▲ *Beer is made from grains of barley and flavored with hops. The barley is first made into malt, and the process of making the malt into beer is called brewing.*

barley

◀ *A head of barley is made up of tiny **kernels**, each enclosed in a sheath or hull. At the tip of the sheath is a long spike known as an awn. In Europe the commonest kinds of barley have two rows of white kernels in hulls. In the United States and Canada six-rowed types are most common.*

Rice, Millet and Sorghum

Rice is the main food of over half the people in the world. In the wild it is a water-loving plant, but when it was first cultivated in Southeast Asia about 4000 years ago it was grown on dry land.

Today "upland" rice is still found in such places, but "lowland" varieties are planted out in flooded fields, called paddies, where they give a far higher yield. As with all the main cereals, many varieties have been developed – there are more than 2,500 different sorts of rice.

Millet

Several kinds of fast-growing, warm country grasses are referred to as millets. They have been cultivated since very early times in the Far East. Some archaeologists think that millet was grown even before rice was used as a crop. The seeds are small and round and are usually eaten boiled like rice, or used to make a sort of gruel. Like most grain crops, some millets are used as fodder for animals.

▲ *Millet is a vital crop to many peasant communities as it can be grown on poor, dry soil and keeps better than many other crops.*

▼ *Rice in the Camargue area of France. "Hi-tech" farmers use laser beams to ensure that the field is entirely flat so that it will flood evenly.*

▶ *There are three main types of rice. Long-grained rice, grown in warm climates, takes longer to ripen but is less starchy. Short-grained rice can be grown in cooler places and is becoming a popular crop in southern Europe. Wild rice is the seed of another sort of water grass. In the past American Indians harvested it from canoes. Nowadays it is grown on a large scale in Canada.*

long-grained rice

short-grained rice

wild rice

common millet

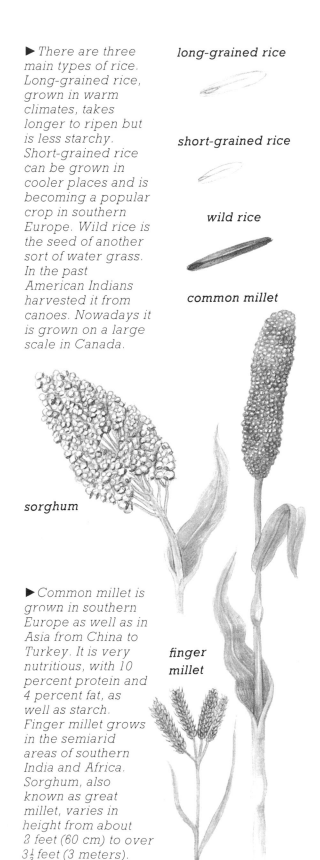

sorghum

finger millet

▶ *Common millet is grown in southern Europe as well as in Asia from China to Turkey. It is very nutritious, with 10 percent protein and 4 percent fat, as well as starch. Finger millet grows in the semiarid areas of southern India and Africa. Sorghum, also known as great millet, varies in height from about 2 feet (60 cm) to over 3⅓ feet (3 meters).*

BIRD FEED

*M*illet is often used for feeding pet birds. Occasionally millet plants will spring up in a garden, or on a waste heap, where bird seed has been thrown away.

Sorghum

Sorghum, which looks like a large millet, grows wild in Ethiopia but is now farmed mainly in Africa, India and China. It is also grown experimentally in parts of Australia. Several hundred varieties are known. Some, grown in southern and eastern Europe, are used as fodder crops. White-grained varieties from more tropical areas are used as human food. They do not contain gluten and so cannot be made into bread, but they are eaten as a sort of porridge. Some varieties are red-grained. These have a bitter flavor, and are used as a basis for beer. At least one of the sorghums has a high sugar content and is grown particularly in the USA, where the crushed stems produce a syrup that is used in cooking.

Maize

Maize is known by many names. In America it is simply called corn. In some areas it is referred to as Indian corn, and in Africa it is known as mealies. It was brought to Europe by Christopher Columbus and is the only cereal crop that has spread from America to other parts of the world.

After wheat and rice, maize is the most important of all grains and is the staple food of people in many tropical countries. It is popular because it will·grow in most warm climates and in most soils. For a given amount of labor maize produces more food than any other cereal.

Cultivated maize

All cereal crops are very different from their wild ancestors, but in maize the difference is so great that nobody knows what its wild ancestor was like. Modern maize could not survive in the wild, for the **cob** is surrounded by leaves that prevent the seeds from being shed. Even if the leaves are stripped away, the central stem, or **rachis**, of the cob is very thick and strong. In wild grasses the rachis always breaks easily so that the grains will scatter when they are ripe. A plant that cannot spread its seeds is doomed.

▼ *In many countries corn is harvested by machine, but in Mauritania it is still picked by hand.*

◀ *The large yellow grains in a ripe cob of maize are set closely around the central stem, or rachis. A sheath, called the husk,* *covers the cob and grains. Maize is an unusual grass in that its male and female flowers are separate but occur on the same plant.*

Uses of maize

Maize is sometimes ground to make a flour that is the basis for many kinds of bread and porridge. The grain is also used to make whisky and industrial alcohol. Corn oil is used for cooking, and the fine ground starch from corn is found in some kinds of cosmetics. In the Western world maize is often used to feed livestock, including pigs, cattle and chickens. About 80 percent of the maize crop of the USA is used for this purpose.

The hunt for wild maize

Archaeologists have hunted for wild maize, but the nearest they have got to it is in some caves in upland Mexico. Here tiny cobs, only about $\frac{4}{5}$ inch (2 cm) long, were eaten by people who lived in about 5,000 BC. It was probably not a very important part of their food, for each cob had only about eight rows of tiny seeds. Even so, it must have been cultivated, for although the rachis was more brittle than that of modern maize, the seeds would not have been spread easily from the cobs. About 3,500 years ago things changed and larger corn cobs were grown.

▼ *Some sweet varieties of maize are harvested before they are fully ripe and eaten as corn on the cob. Maize grains can be* *made into cornflakes and one kind of maize with a high water content is used to make popcorn.*

Vegetables

A lthough cereal crops are more important to our diet, vegetables add variety and are rich in salts and vitamins. They have been cultivated for many centuries.

Various parts of vegetables are eaten. Sometimes it is the root where plant food is stored. Other vegetables, such as the potato, have underground stems that become swollen with food. Peas and beans are examples of seeds that are eaten as vegetables.

In some plants, such as celery and kohlrabi, the stems of the leaves are eaten. Rhubarb and asparagus are further examples of stem crops.

Leaf vegetables

The leaves of plants are often too strongly flavored to be pleasant to eat. Even so, they contain vital vitamins and minerals and without them people develop deficiency diseases. A little seaside plant called scurvy grass used to be gathered

▲ Celery can be grown in trenches, so that the stems stay white by being covered with earth.

▼ A selection of vegetables on a market stall.

by sailors, who ate its sharp-flavored leaves to combat **scurvy**. This awful disease was common on ships where the food consisted mainly of salt meat and biscuits, with no fresh fruit or vegetables.

Spinach is probably the most nutritious of all the leaf vegetables. It came originally from Iran and did not reach Europe until the Arab conquest of Spain. It was not grown on a large scale until the 18th century, and even now is often a market garden rather than a farm crop. Several other relatives of spinach, such as fat hen and good-king-henry, were cultivated in earlier centuries but they are now regarded as weeds.

Today the cabbage family is one of the most important providers of food leaves. Wild cabbage is an edible but strongly-flavored plant that grows along seashores in western and southern Europe. It has been cultivated by farmers for at least 4,000 years, and many varieties have developed. In true cabbages and kale the leaves are used. In cauliflowers and broccolis the unopened flowers are cropped, while in brussels sprouts, buds grow into tiny cabbages up a main stem.

Onions

The strong flavor of onions and similar plants has been used to enliven food since biblical times. It is not certain what the wild ancestor of the common onion was like, or where it came from, but it was probably native to Turkey. Today it is eaten all over the world.

Onions take two years to mature, storing the food made in the first year in a **bulb**. They may be grown from seeds, or from setts – which are small bulbs, harvested after one season's growth and replanted the following spring.

Onions are eaten in many ways, sometimes they are used as a vegetable in their own right and sometimes as a flavoring. Usually the bulb of the plant is eaten but in some cases, such as chives, the leaves are also used.

▼ *Shallots, leeks and garlic are all related to the onion but the food storage organs are differently shaped.*

leeks

onion

shallots

garlic

Root Crops

Some scientists think that when human beings started to farm, their first crops were not cereals, but tubers and roots. These are often easier to grow than grain and if a little bit of root is left in the ground it will make a new plant for next year's crop.

Most root crops are rich in sugars and make valuable foods for humans and livestock. Turnips and mangels are commonly used for feeding cattle and horses. The residue of sugar beets, after the removal of most of the sugar, is made into pellets for animal foods.

Carrots

The wild carrot is a common European plant that was cultivated by Iron Age farmers in southern Europe and is now grown in many parts of the world. It has many uses, for it is a good source of

◀ *Sweet potatoes are an important food in the tropics, for they only grow in warm, wet climates. Their tubers are often pink skinned, and have yellow flesh. The sweet taste comes from the sugar that they contain, but they are also a useful source of starch and a little protein.*

▶ *Yam tubers are harvested about eight months after planting. A single tuber may weigh up to 110 lbs (50 kg).*

vitamin A and the seeds produce an oil that is used in making perfumes and for flavoring liqueurs.

Manioc

In the tropics many kinds of roots are cultivated. The roots of the manioc or cassava shrub can be eaten in various ways, fermented to make beer, or fed to livestock. In areas where the soil is poor it is the best source of starchy food. There are many varieties, some of which contain a high proportion of poison. This is removed by cooking or drying in the sun.

Yam

Yams are the main food for millions of people. The plant is a vine that grows well in warm, wet places of the world. Unlike many other root crops the starch and sugar cannot be removed to make a cash crop, but yam tubers contain chemicals that are used to make medicines.

Sweet potato

This American plant is an important source of sugars and vitamins and is used as well in the production of starch, glucose and alcohol. It is also grown in the warmer parts of Japan and China. Sweet potatoes are sometimes called yams, but they are not true yams.

POTATOES

*A*lthough not strictly root crops, in the sense that they are not swollen roots, potatoes are the main vegetable that grows under ground. They were grown on a small scale by Peruvian Indians high in the Andes, and were brought to Europe in the early 1660s. They were taken to North America more than a century later. Today they are farmed as far north as Alaska and southern Greenland, but they grow almost everywhere except in the lowland tropics.

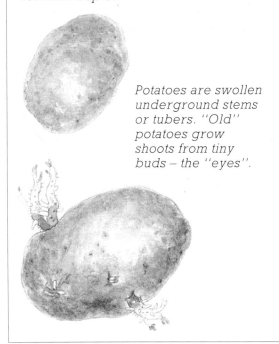

Potatoes are swollen underground stems or tubers. "Old" potatoes grow shoots from tiny buds – the "eyes".

▼*In Brazil manioc is grown to make sugars, tapioca, glue and acetone.*

Peas and Beans

Peas and beans belong to a huge family of plants found almost all over the world. Their seeds grow in pods and are easy to harvest. Some members of the family are large trees, but these are not important as providers of food.

Peas and beans are important to farmers because their roots carry small nodules that release **nitrogen**, important to plant growth, into the soil. So they actually improve the soil that they grow in. Small members of the family, including clovers and similar plants, are often grown with grasses for this purpose.

Pulses

In early times wandering tribes probably found that the seedpods of vetches and similar plants were good to eat. Later they discovered that the seeds could be dried and when soaked in water would become soft and edible again, so they could be kept as food for the winter. For much of human history peas and beans have been

▲ *Beans being harvested for freezing. Nowadays many growers of peas and beans are contracted to processing firms, which take the crop and freeze it within two or three hours of picking.*

used in this way. Dried peas, beans, lentils and similar seeds are known as **pulses**.

The origin of peas

Peas of a kind were cultivated in Turkey more than 8,000 years ago. Garden peas similar to those of today were in use before Roman times. Today several species are grown, including cowpeas, which are eaten in the tropics. In this species the leaves as well as the pods or the dried

FOOD FOR VEGETARIANS

*B*oth peas and beans have a high food value. Some beans contain up to 25 percent protein and 64 percent sugars and starches, so they are especially useful to people who do not eat meat. Peanuts are also extremely nutritious, containing 30 percent protein and up to 50 percent oil. They are rich in vitamins B and E. The nuts are produced under ground and have to be dug, like root crops.

world until the 20th century. Although the seeds are sometimes eaten as green vegetables, they are mostly used to make oil and meal. In the USA soya beans are used to make oils, margarine and many manufactured goods. The meal is mainly useful for making concentrated food for farm animals.

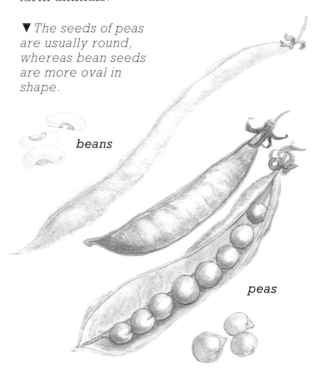

▼ *The seeds of peas are usually round, whereas bean seeds are more oval in shape.*

beans

peas

seeds are eaten. Some varieties, known as yard-long beans, have pods up to 40 inches (1 meter) in length.

Beans, beans, beans . . .

Broad beans may have come from western Asia, but they have been grown in Europe since prehistoric times. More kinds of beans come from America, where 84 species are known. Four of them, including runner beans and lima beans, are important crops.

Soya beans

The soya bean comes from the Far East. It has been grown in China since ancient times, but did not reach the Western

▼ *A pea plant with ripe pods. The plant climbs by means of tendrils because the stalk is too weak to stand upright.*

Fruits of Cool Lands

A plant's seeds are held in its fruit. This may be hard and dry, or fleshy and succulent, though most people think of fruits as being sweet and juicy, like apples.

Apples are the most important fruit grown in cooler parts of the world. They have been cultivated for at least 2,000 years in Europe. They are used for eating, cooking and making cider, and there are many varieties for each of these purposes. Apples do not **breed true**. So if you grow an apple pip of a variety that you enjoy, you will find that when your new tree bears fruit it will be nothing like its parent. It will probably taste like cotton balls wrapped in cardboard, but very rarely a well-flavored fruit occurs. This is maintained by **grafting** buds onto strong-growing crab apple stock.

Plums and sweet cherries are other orchard fruit that have been grown in Europe since very early times. Pears, which came first from Central Asia, had reached Europe by 1000 BC. These fruits do not keep as well as apples, so until recently they were far more seasonal. The quince, which came from western Asia, is grown commercially in Spain and Portugal, but its main use elsewhere is as a **rootstock** on which pears are grafted.

Soft fruits

Soft fruits include the many kinds of brambles that are native to the cool northern parts of the world. Cultivated blackberries are derived from an American species. More popular and more widely

◄Modern apple growers prefer miniature trees, which are easier to pick and to deal with in the yearly routine of pruning and spraying needed to maintain a high yield.

grown are raspberries, and the logan-berry, which is a natural hybrid between blackberries and raspberries. In Europe native strawberries have been cultivated for many centuries, but the fruit of these plants is small. Modern large strawber-ries are a hybrid of two American species. As with raspberries the fruit is eaten fresh, or used for jams, flavorings and ice creams.

The thin, acid soils of mountains and moorlands of the far north produce some valuable fruit. The cloudberry is one of the most delicious. It is rarely cultivated but is collected and sometimes used to make jams and jellies. Bilberries and cranberries, relatives of the heather, are not often cultivated, but the wild fruit are gathered for making jellies and jams and sauces to eat with meat. The American cranberry is now grown on a large scale in some parts of Europe and its much larger fruit has replaced the small native cranberry that is far more difficult and laborious to collect.

▼ *The crab apple has small sour fruits. Breeding and selection from it has yielded thousands of varieties of cider, cooking and eating apples.*

▲*Grafting is used to increase the numbers of woody plants that do not breed true, or are difficult to grow from seed. Many plants that we buy, like roses and fruit trees, are grafted.*

To make a graft, a gardener cuts a small piece, the scion, which must contain a bud, of the plant that is wanted (1). This is attached to a rooted plant of a strong growing, closely related kind, which is called the rootstock (2). The joint is protected so that sap is not lost (3). If the graft is successful, the scion will form a new plant on the rootstock.

Fruits of Warmer Lands

A hundred years ago most people could eat only locally grown fresh fruits at the time that they ripened. Nowadays, shops and markets sell fruit throughout the year because crops grown all over the world can be refrigerated and transported. Many come from warm, but not tropical, countries.

Fruits of the orange, or citrus, family are among the most important. The citron was known in Europe in Roman times, and bitter oranges were brought west by Arab travelers about 1,000 years after this. Arabs also brought lemons (which are unknown in the wild) in the Middle Ages, but sweet oranges were not traded until the 16th century and reached America shortly after this. Other citrus fruits such as grapefruit, ortaniques and tangelos are probably hybrids. Sweet oranges and lemons are now grown in Mediterranean countries and places with a similar cli-

mate. The USA and Brazil are the biggest producers, with Spain, Italy, South Africa and Australia also important.

Grapes

There are over 60 species of grapes growing wild in North America and Asia. Many are edible, but one kind that came from western Asia is more important than all of the others. It reached the Mediterranean lands at least 4,000 years ago and was well-known to the Ancient Egyptians, the Greeks and the Romans. Grapes grow

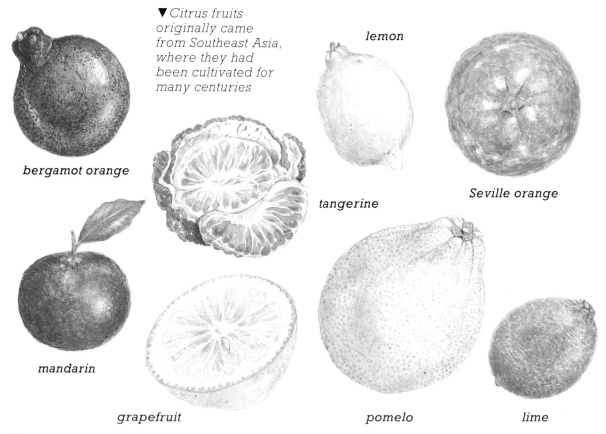

▼ *Citrus fruits originally came from Southeast Asia, where they had been cultivated for many centuries*

lemon

bergamot orange

tangerine

Seville orange

mandarin

grapefruit

pomelo

lime

WINE

*M*ost grapes are used for making wine. While wine is not a food, many people feel that it adds to the quality of life. The mysteries of wine making and its effect on those who drink it became an important part of many religions. The Romans took vines to much of their empire, and in the warmer zones of this great area from western Asia to France and Spain and North Africa, vines are still grown. France is the main wine producer, but vines have now been taken to North and South America, South Africa and Australia, and all of these areas produce their own wine.

on vines that are carefully tended, grown and pruned according to the crop required. Some are eaten as fruit, and others, mainly seedless varieties, are dried to make raisins, sultanas and currants.

Tomatoes

Another plant that needs warmth to ripen its fruit is the tomato. Wild tomatoes with small, cherry-like fruit were grown in Mexico at the time of the first European exploration. Seeds were taken from there to Spain and Italy, but although new and larger varieties were developed, tomatoes did not become a popular plant until the end of the 19th century.

◀*Dates will not thrive in wet climates, but have been grown in dry, warm areas of the Middle East for at least 5,000 years.*

▼ *Tomatoes are grown all over the world, but apart from places with really warm summers, most varieties are greenhouse crops.*

Tropical Fruits

Far more kinds of plants grow in tropical climates than in the cooler parts of the world. Many of them produce fruits that are eaten locally. Some of these are delicious, but they are rarely seen outside their country of origin.

▲Bananas do not grow on woody trees, but on plants made of huge overlapping leaves. When the plant is mature, a bud emerges from the leaf crown and produces a many flowered cluster. There may be as many as 200 bananas on a single bunch. The cultivated fruits are sterile, so new plants are produced from suckers or buds.

Some such as the small ladies' fingers bananas do not travel well. Other fruit grows irregularly or in small quantities in any one place, so that commercial cropping would be very difficult.

Pineapples

The pineapple was discovered by early travelers to the West Indies. By 1600 it had been taken to may other parts of the world and it is now farmed on a large scale. Hawaii produces more than anywhere else, followed by Brazil. The plants are not grown from seed, but from **suckers** or **crowns**. These can produce fruit for many years, but are usually replaced after three harvests, because the fruit gets smaller after each crop is taken.

Bananas

Bananas need warmth and a humid climate if they are to thrive. They came originally from Southeast Asia, where they have been a popular food for thousands of years. They spread to India and Africa, and were taken to America during the 16th century but trade in them did not become important until late in the last century. Brazil and Equador are the main exporting countries. Bananas are cut green for export and ripened in store.

Mangoes

Mangoes grow on large, true trees. The sweet and juicy fruit has, at its best, such a fine flavor that it has been cultivated since early days and even has a place in the Hindu religion. The fruit can be eaten raw or preserved in various ways, especially as pickle or chutney. One problem is that the fruits are very fragile and do not travel well.

Pawpaws

Pawpaws, which are also found throughout the tropics, came first from South America and the West Indies. Like many tropical fruits, pawpaws grow directly from the trunk of the small trees that bear them. As well as being delicious the fruits are said to have medicinal properties. The tree is used to produce a gum with many uses from making medicines to shrink-resisting woollens!

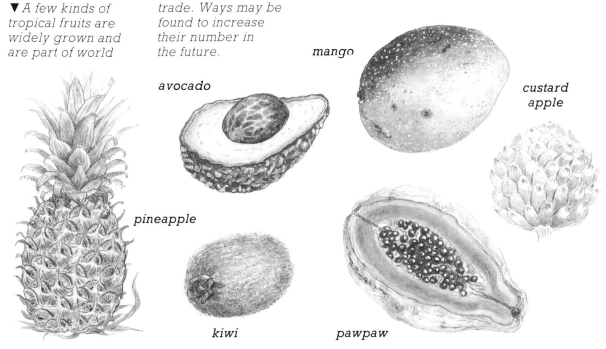

▼ A few kinds of tropical fruits are widely grown and are part of world trade. Ways may be found to increase their number in the future.

mango

avocado

custard apple

pineapple

kiwi

pawpaw

◄ Breadfruit grow on large and beautiful trees that are native to Malaysia and Southeast Asia. They are now found all through the tropics. The purpose of Captain Bligh's famous voyage on the Bounty was to take breadfruit seedlings from the Pacific to the West Indies. The oval green fruits are about 8 inches (20 cm) in diameter and consist of 30–40 percent carbohydrate.

Nuts

Nuts are large, oily seeds, protected by a hard but brittle shell. They are mostly produced by trees, but we also use the word for certain hard root structures, such as tiger nuts, which are made by a sedge, or earth nuts, which are part of the roots of a plant related to parsley.

Nuts are not only very nutritious; they keep well and can be stored easily. Human beings have probably always eaten nuts. The remains of large numbers of pistachio nuts have been found in the excavation of a settlement dated about 7000 BC in southern Turkey. It is known that even earlier than this people in the southwestern United States depended to a large extent on nuts that they collected.

As food, nuts are rich in oil and protein, but they are often very labor-intensive to collect and prepare, so they tend to be

▲ Nuts are eaten in many ways. They are the basis of many vegetarian foods. Salted, they are often used as snacks, and they are also important in confectionary.

◄ In Brazil the babassu nut is broken open and the heart is taken out. It is used to make oil and an alternative source of fuel.

expensive. While there were great forests covering much of the earth, most nuts were collected from the trees as they grew naturally, and this is still the case to some extent. Brazil nuts, for instance, are mainly picked up from the forest floor. Some nuts, like acorns and beech nuts, are good food for some livestock.

▼ *A selection of edible nuts with their shells.*

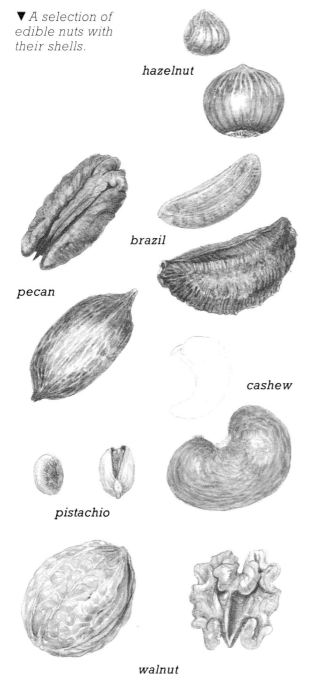

hazelnut

brazil

pecan

cashew

pistachio

walnut

Hazelnuts

Hazelnuts still grow wild in much of northern Europe. Their bright yellow "lambstail" catkins are a sign of spring, and in autumn children and family parties still collect the small, sweet nuts. Hazelnuts are also cultivated, especially in Turkey, Spain, Italy and Oregon. Barcelona nuts and filberts are two widely grown varieties. As well as being eaten fresh, the nuts are sometimes crushed for oil, which has many uses.

Walnut

Walnut trees came originally from southeastern Europe and western Asia, but the value of the nuts as food and for oil has meant that they have been planted throughout much of the world. The United States now produces more than Europe, where France is the chief grower. American species of walnut, the black walnut and the butternut, are collected and are important locally. The pecan, which belongs to the same family, is grown in the south for its thin-shelled nuts, some of which are exported. The trees are also valued for their wood.

Pistachio

Pistachio nuts grow in the dry parts of southeastern Europe and the Middle East and in the United States. The trees are small and slow-growing, though they continue to fruit for many years.

Macadamia

The macadamia nut grows in tropical climates. It is sometimes called the Queensland nut, for it grows wild in northeastern Australia, but it is now cultivated in a few places, particularly Hawaii. The nuts are small and round and encased in a very hard shell. It is among the most delicious of all the nuts, though dieters should avoid it, as the fat content is about 70 percent.

Oil from Plants

All of the nuts contain a great deal of fat and oil. Sometimes this is extracted and may be used mainly in cooking, as with walnut oil, or cosmetics, as with almond oil. But oils have many other important uses, including making soap, paints, and industrial lubricants.

The most important plants for the extraction of oil are the coconut palm and the oil palm. These are both found only in the tropics, where they are a vital part of the economies of many people.

Oil palm

Oil palm comes from West Africa, but plantations have been established in other parts of the continent as well as in Malaysia and Indonesia. Every year mature palm-oil trees produce several bunches of fruit, each weighing up to 32 lbs (15 kg). Two varieties of oil are obtained. One, palm oil, comes from the outer, fibrous part. A good deal of palm oil is eaten in Africa. When exported it is mainly used for making soap, or for industrial purposes. The finer palm-kernel oil is prepared from the inner nut, and is mainly used for making margarine.

Other oil plants

We get oil from other plants, including the olive, which gives a high quality product used as a salad oil and for cooking. Peanut oil is specially good for frying, as it does not alter with heat. Sunflower oil is also used for cooking. It is polyunsaturated, and so is less likely to cause a buildup of

▶ *The white flesh of coconuts is dried to make* copra. *This gives coconut oil, which is used in soap-making, as cooking oil and in margarine. As with palm oil, the residue of the nut, after the oil has been extracted, is valuable as feed for livestock.*

coconut palm

oil palm

▲ The olive tree grows in the Mediterranean area and in California. When ripe, the black fruits contain up to 60 percent oil, which is extracted by crushing and pressing. The oil is used in salads and for cooking.

cholesterol in the arteries. Oilseed rape, a relative of the turnip, is also widely grown. In the past the oil from this plant was used in lamps, but now it is in demand as an industrial lubricant, though some goes to make margarine. One of the most important oils is that from the soya bean. It is the chief vegetable oil produced in America and is not only used for cooking but in making paints, synthetic rubber, inks, soaps and many other products.

Jojoba oil

In the future vastly increased areas may be given over to the cultivation of the plant that provides jojoba oil. This comes from dry areas in North America and so it would be a good crop for other arid places. The oil has been known for a long time, but recently it has been discovered that it can replace many of the industrial uses for which whale oil was thought to be necessary. It may be that the use of jojoba oil could save some species of whales from extinction.

soya

▼ Vegetable oils are made from soya beans, sunflower seeds, peanuts and rape.

rape

peanut

sunflower

OTHER SOURCES OF OIL

*O*il can be extracted in small quantities from many other plants. Even some waste products, such as grape pips, give a high quality salad oil, and plants such as juniper and the clove tree, among many others, produce an oil that is useful in medicine.

The Sweeteners

All green plants contain sugars that they may store in roots, stems or even flowers. In moderate amounts sugar is good to eat, for its energy can be released quickly. Unfortunately, too much sugar can lead to many problems, from tooth decay to obesity.

These troubles did not bother the human beings who lived in Southeast Asia about 10,000 years ago. They found a source of sweetness in a large grass, which was pleasant to chew, that we now call sugarcane. It is thought that sugarcane was first grown in New Guinea, and from there was taken to India in very early times.

In Europe the Romans knew of sugarcane but it was rarely traded so far west and their cakes and candies were sweetened with honey. Sugarcane was treated as an uncommon spice to be chewed. Later it was grown in the Mediterranean area, and was taken to America within a century of Christopher Columbus's

▲ *Sugarcane is almost always grown as a plantation crop, for the sugar is extracted by putting the cane through heavy rollers and concentrating the juice by boiling it, something that a single farmer, working on his own, cannot do.*

▶ *A small hole is drilled into the trunk of the maple tree and the sap is collected as it flows.*

discovery of the continent. Even so, sugar did not become an important part of the western diet until the early 18th century.

Nowadays, high yielding forms of sugarcane have been developed. These produce more human food per acre than any other crop. It is grown throughout the tropics, and many small countries rely on it above all else for their trade.

Sugar beet

Sugar beet is grown in cool parts of the world. Its relatives, sea beet and red beetroot, had been eaten for many years, but in 1802 the sugar in sugar beet was discovered by a German chemist. Napolean encouraged the growth of sugar beet when Europe was blockaded during the Napoleonic wars, but after that its use dwindled. Now huge amounts are grown, especially in Russia and Europe.

Maple syrup

The sugary **sap** of several kinds of trees is tapped to give sweet liquids. Maple syrup comes from the North American sugar maple tree. Its sugar content may be as much as 11 percent; it is concentrated to over 66 percent by boiling.

In the tropics the wild date, or toddy palm, is tapped in a similar way. A large tree may give as much as 350 lbs (160 kg) of sap, which can be refined to about 35 lbs (16 kg) of sugar.

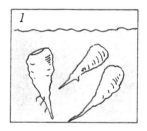
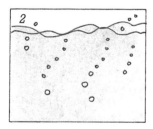
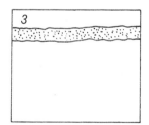

1. The roots of sugar beet are soaked.
2. After treatment the liquid is boiled to produce a syrup.
3. The sugar crystallizes out.
1. The tops, which are cut off before harvesting, are fed to livestock, as is some of the pulp left after the sugar has been removed.
5. A use that may become more important in the future is the production of ethanol, a fuel, from partly purified sugar solutions.
6. All these things contain sugar in various forms.

Plants as Fuel

Human beings have known how to make fire for at least 750,000 years. The chief fuel that they have used has been wood. Today, the use of wood as fuel continues, but the great forests that existed in early times have in many areas been destroyed and the demand for wood is far greater than the supply.

◄ Coal is the fossilized remains of forests that were alive 345 million years ago. Fossilized leaves are often found in pieces of coal.

▲ Tropical forests like this one in Borneo are being cut down both for fuel and to provide land for farming. Without the plants to protect it, the soil is washed away and the land becomes barren and useless.

In the foothills of the Himalayas the forests have been cut down by peasants who need fuel, mainly for cooking. They have since discovered that the growing trees were of more use to them than they suspected, for once the forests have gone, there is nothing to hold the soil, which is swept away in the monsoon floods.

In cooler parts of the world, the greater forests were destroyed many centuries ago. Here, there has been less damage to the environment, for the climate is less harsh. Even so, it is difficult to imagine, when in the farming country of southern Britain, that this was, until 150 years ago, that nation's industrial heartland. Iron was smelted and steel made, using oak trees from local forests as the fuel.

Today, wood is still used on a small scale to heat houses, though great open fireplaces burning huge logs are a rarity. Instead, much more efficient closed stoves are used. These have the advantage that almost any sort of wood can be used in them, even kinds that would smoulder and spit on an open fire.

Coal

On a worldwide scale wood as a fuel was replaced by coal. Yet this also is wood, for it is the remains of trees that formed swampy forests many millions of years ago. In some places the decaying timber hardened to make coal. When we burn coal today we are using the energy accumulated in the distant past.

New sources of power

People are looking for new sources of power. One of them, already in use, is alcohol made from plant sugars. Brazil is the pioneer in this, using sugarcane as the raw material. Alcohol is obtained from molasses after refined sugar has been removed. Cars can be run on this fuel, though the engines have to be modified.

In the USA a desert plant, the buffalo gourd, has been shown to produce an oil that can be used instead of diesel fuel.

PEKING MAN

People who lived in caves near Peking more than 300,000 years ago discovered how to feed and control fire, so that it could be useful to them. Their fires were fed with dead branches, for they did not have tools suitable for felling trees. But fire was valuable, for it enabled them to have light at night, to keep warm, to frighten away wild animals and most important of all, to cook their food. They may have used Aleurites as fuel as its wood is very oily and burns well.

Plants for Building

Since the early Stone Age people have used plants to make shelters. At first these were probably just dead branches covered with a thatch of small twigs or grass, but as soon as they had invented axes and adzes they could build real houses.

Wood is still a very important building material. In parts of the world where forests survive, houses are built entirely of wood. Elsewhere, wood is used as rafters to support the roof, and to make doors and window frames.

Hardwoods

Trees vary in the size and strength of the wood that can be obtained from them. In Europe, oak was at one time the most important timber. As the local oak forests were destroyed, tropical timbers were imported to replace them in building and for other uses.

In many areas the valuable **hardwood** trees have been removed from these forests as well. In some cases attempts

▲ *Axes were used to cut down trees, and adzes to work the wood into planks.*

◄ *A modern wooden house. The walls are constructed of planks of wood. The shingles on the roof are made of Western red cedar, an American species that is often used for this purpose because the wood is not heavy and therefore does not need very strong roof supports.*

have been made to grow plantations of useful species. Teak, for instance, which is native to Southeast Asia, has been planted in the West Indies and Africa for it is one of the strongest and hardest of woods.

Softwoods

But hardwood trees are slow growing and it takes many years before a crop of wood can be harvested. As a result, there has been an increased use of **softwoods** – the timber produced by northern coniferous trees such as pine and spruce. Generally this grows faster than other kinds of wood, and in some areas plantations of foreign softwoods have replaced the old forests.

Roofing materials

In the past roofs were sometimes made of wood, but more often were covered with a thick layer of grass or reed stems, which formed a thatch. This makes a good insulation, but needs to be renewed more often than other sorts of roofing. In areas where clay for bricks and tiles is not available, tiles or **shingles** of wood may be used.

NAMING OF PLANTS

*T*he scientific names of plants are often difficult to understand, but they usually have a meaning. One that has a bearing on wooden buildings is a moss with the cumbersome name of Fontinalis antipyretica. Fontinalis *means coming from the water, and* antipyretica *means against or rivaling fire. This refers to the fact that wooden houses of the 18th century often had the cracks in the boards stuffed with moss. This particular species was chosen because it would not catch fire, so the dry wood was protected by the moss.*

▼ *The base and frame of a wattle-and-daub house were constructed with huge oak timbers.*

The spaces in between were filled with small woven branches, or wattle.

The wattle *was then covered with a mixture of mud and other substances, called* daub.

The walls were then plastered over and whitewashed.

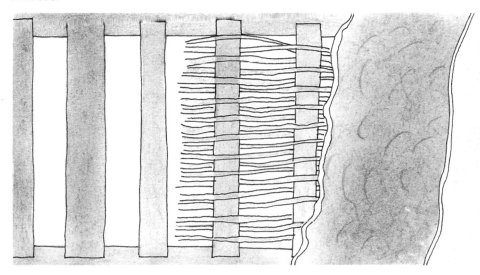

Wood for Furniture

Many kinds of trees produce wood that is not large or strong enough for the main timbers of buildings. Even so, they may be useful in other ways, for example in making furniture.

Very many species of trees have been used for this, even bamboos, which are used for cane furniture. In general, people want to surround themselves with beautiful objects, so the woods most favored for furniture-making were those with a decorative grain, which could be carved easily and polished well. At least five species of tropical trees are known as satinwood because of this.

Walnut

In Europe, oak was the wood chiefly used in making furniture until about the 16th century. After this time, fine and elegant

▲ *Nowadays, many pieces of furniture showing a fine grain or pattern of wood are in fact not solid but veneered. In the past, veneers were often used to make the elaborate patterns that decorated pieces of furniture and buildings.*

◄ *Present-day mahoganies come from the mainland of America, from Mexico to Brazil. Some African trees are also known as mahogany, but the chief of these, a forest giant called Khaya ivorensis, is now threatened because of over-exploitation.*

furniture was made of walnut, which has a beautiful grain and takes a very high polish. In America different species of walnut were used, particularly the black walnut and great numbers of these trees were cut down.

Mahogany

Partly because of the increasing rarity of walnut, and partly because of changes in fashion, the most popular wood of the 19th century was mahogany. This is a reddish-colored wood with a fine grain. It can be carved and polished easily. Originally mahogany came from the West Indies, but the island forests were largely cut down and destroyed soon after their discovery by Europeans.

Veneers

The cost of fine woods has always been high, so since the days of the ancient Egyptians and Romans furniture makers have taken very thin slices of rare or especially beautiful wood and stuck them onto more commonplace and cheaper bases. The thin slices of wood are known as **veneers**.

Plywood and chipboard

Today many wooden articles, including furniture, are made of **plywood**. Plywood is several thicknesses of wood glued together in such a way that the grain of each slice is at an angle to those on either side of it. This gives the plywood much greater strength than solid wood. Often a weak but beautiful species is used for the face wood, so modern furniture uses far more kinds of woods than was possible in earlier times.

For some uses, such as built-in cupboards, even cheaper wood is used to make **chipboard**. This is formed of small fragments of softwood, bonded with a resin. It is often finished with a highly polished plastic surface, made to look like a beautiful wood.

THE SPINDLE TREE

Sometimes the name of a tree gives a clue as to its use. The spindle tree, for instance, is not large, but it has a hard, fine-grained wood that was turned and used for making spindles, which had to take a lot of wear, but not a great deal of weight.

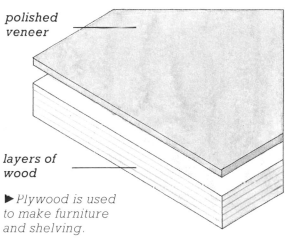

polished veneer

layers of wood

▶ *Plywood is used to make furniture and shelving.*

Paper

Books and birthday cards, computer printouts and chocolate boxes, food sacks and newspapers have one thing in common – they are all made of paper. Paper is so much part of our world that we take it from granted, yet it is another example of our dependence on plants, for it is made of plant fibers.

About 2,000 years ago the Chinese invented a way of pressing fibers of various kinds together to make paper. They used bamboo, rags and other materials. The process was kept secret for many centuries, but gradually spread to the Western world. Until the last century rags were used to make paper, although there were experiments with other materials, including moss and wasps' nests.

Birth of the paper industry

In 1719 a French scientist suggested that wood pulp could be turned into paper, but he was ignored. In 1843 a German discovered how to make paper using wood pulp and the modern paper industry, which is the world's largest consumer of wood, was born.

Most paper is now made from coniferous wood, especially fir and spruce, which grows in the northern parts of the world. Other kinds of trees are also being used. These include quick-growing species such as birch and poplar, and in the southern parts of the USA a kind of hibiscus is being grown for use in pulp mills. Many agricultural waste materials, including straw and sugarcane leaves, may be used more in the future.

Bamboo

In the tropics, other woods are used for papermaking. Bamboo is one of the most important, for this has fibers that are long and soft. On their own they make a high grade paper, but are often mixed with other materials for a cheaper product.

▲ *So great is our need for paper that huge areas of forest are taken each year. One thousand acres (400 ha) of forest are needed for a single press run of the* New York Times. *In time, the trees grow again, and some areas are logged several times. But no resource can stand such heavy use. Now the logging companies are replanting as they cut, so that there will still be forests in the future.*

HOW PAPER IS MADE

1. Felled trees are taken to a pulp mill, where the wood is shredded into chips.

2. Other fibers such as straw or grass may be added.

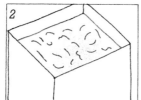

3. The fibers are heated with water and chemicals, which reduces them to pulp.

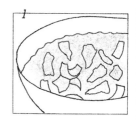

4. The pulp is washed and bleached. Bales of pulp are sent to a paper mill, where it may be dyed, or have other substances added, including fillers to give the paper weight.

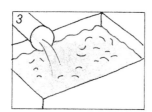

5. The liquid paper flows onto a mesh belt, where the water drains away.

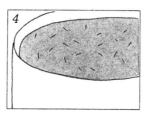

6. The remaining felt-like material is passed between hot rollers, which flatten and dry it.

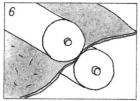

Rice paper

Other special papers may be made from particular plants. Rice paper, used in the Far East for painting, making toys and artificial flowers, comes from the pith of a plant called *Fatsia papyrifera*. Small and specialized uses like this have little bearing on the great problem of how to maintain raw materials for our growing need for paper. Probably the final answer will be in more recycling.

MAKE YOUR OWN PAPER

1. Tear some newspaper into small pieces. Soak in water until reduced to a pulp.

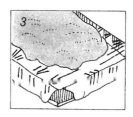

2. Squeeze out as much water as possible.

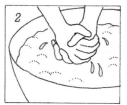

3. Spread the pulp out in a thin layer to drain. (A piece of nylon stretched over a baking pan makes a suitable draining rack.)

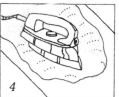

4. Use a cool iron to flatten the dried-out pulp into a smooth sheet of paper.

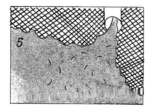

Fibers

We do not know when human beings first began to use plant fibers. It is possible that Old Stone Age people first used them for making ropes. The strong stems of climbing plants, such as old-man's-beard, would have been ideal for tying bundles of firewood, or dragging the heavy carcass of an animal. Later, strips of bark were twisted together to make a more flexible rope.

In West Africa the bark of the tree *Cleistopholis patens* is still used in this way. Before the end of the Old Stone Age, people began to weave baskets of pliable twigs. As time went on, finer materials were used. Today, some North American Indian tribes still have the secret of making baskets with grasses so finely woven that they will hold water.

Twigs and grasses are too coarse to weave into a soft, wearable fabric. But some plants, like cotton, have long hairs attached to their seeds, and in a few cases these can be spun or twisted together to make a thread from which cloth can be

▲ *Flax has been used since ancient times. It grows in cool, moist climates in Europe and Asia. To get the long, tough fibers the plant is cut and* **retted***, or soaked in water to get rid of most of the soft tissue. It is then scutched or beaten, combed, carded and spun into linen.*

◄ *Cotton growing in Texas. Most of the cotton in the USA is now harvested by machine, but in countries like India and Egypt it is still done by hand.*

woven. Other plants such as jute and flax have long tough fibers that can be used to make cloth.

Jute

Although linen from flax is a fine fabric, another long-fiber plant is more important in terms of world trade. This is jute, which is used mainly to make sacks and matting. Like flax, the jute has to be retted, but it grows quickly, so that the same land can produce two or three crops in a single year. Jute comes mainly from India and Bangladesh, but other warm countries, including China and Brazil, are now growing it.

Other plants that yield fibers are grown on a smaller scale. They include hemp, which is used mainly for making ropes and heavy cloth.

China grass

In the past, stinging nettles were grown for the long and very fine fibers that could be obtained from them, but as you can imagine, they were unpleasant to handle. Until recently some American Indians used wild stinging nettles for making ropes and fishing nets. A relative of stinging nettles, called China grass, produces fibers that are sometimes over 16 inches (40 cm) long – longer than those of any other plant. It gets its name because it was known in China more than 4,000 years ago. Today it is grown mainly in China and in Florida. The fibers are at first sticky and difficult to deal with. But perhaps they are worth the trouble, for their strength is seven times that of silk.

▼ Although clothes are now often made from artificial fibers such as nylon, many fiber plants are still cultivated. Some of the main ones are shown below.

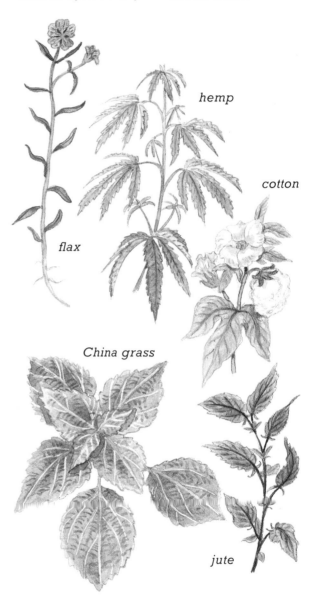

hemp

cotton

flax

China grass

jute

◄ Cotton fibers, which are up to 1 inch (2.5 cm) long, were among the first to be cultivated. Cotton was grown in India more than 5,000 years ago, and slightly later in Peru. Today, cotton is farmed almost everywhere that has a suitable climate. It is an annual plant that needs an average temperature of 59–63°F (15–17°C) during the growing season.

Pigments

People often use the cheerful colors of flowers to brighten their homes and their surroundings. But dyes that can be used to stain fabrics or other objects may be obtained from many plants, and are often more vivid than any flowers. Some of the substances that are used tan and strengthen fabrics, as well as color them.

Plant dyes are often tricky to use, for the colors may fade without a mordant or chemical wash before the dye. It is also difficult to get an exact match of a color, since the intensity of the dye may vary. As a result, natural dyes have tended to drop out of use, and are replaced by chemical aniline dyes. Even so, in the Western world hand spinners and weavers still like to make their own dyes. In many parts of the tropics the richness of natural dyes has meant that they are still used, although generally on a small scale.

Woad and Indigo

The most famous plant dye is probably woad, said to have been used by the ancient Britons to paint their bodies and later to dye fabrics. The blue color is obtained from the pounded leaves. Woad may be grown as a curiosity, but is not now important as a dyestuff.

Another blue dye comes from the indigo plant. This grows in the Far East and was once exported. As in most plant dyes the plant was cut up and soaked in water to release the pigment.

Henna

Another dye that has survived in use is henna. The shrub, whose leaves produce an orangey-brown dye, has been grown in North Africa, the Middle East and India since early times. It is used mainly for coloring hair but it is also sometimes used for fingernails.

▶ *Dyes may be obtained from almost all parts of plants. In some cases the roots are used, in others the stems, leaves, sap, fruit or seeds provide the color. The wood and bark of some species are important.*

woad

henna

indigo

◄ The yellow dye gamboge, from Southeast Asia, is obtained by tapping the sap of the tree. This is allowed to harden before use. It gives a brilliant yellow dye that is used not only to color the robes of Buddhist monks, but also in fine quality paints and inks. Most of the clothes made in Asia use natural dyes.

▼ In Iceland, Norway and Sweden, lichens were used to dye wool.

Annatto and Sanderswood

Annatto is another tropical dye. It comes from the seeds of a small South American tree and was at one time used to dye silk and cotton fabrics. Unlike many such products, which are now used on a very small scale, annatto survives and is even grown as a crop in India. This is because it is nonpoisonous and so it is used to add color to food, especially sweets, butter and margarine.

Sanderswood produces a deep red dye used by Hindus to make caste marks. It is soluble in alcohol and so is used to color some alcoholic drinks.

annatto

sanderswood

Gums and Rubber

If a plant is cut or injured, it will often produce a flow of sap. Sometimes, this is not just a watery fluid, but is sticky or rubbery. Human beings have known this since at least the times of the early Egyptians, who used natural gums in the process of embalming their dead. Since then, people have made use of gums and rubbers in a tremendous number of ways.

Some sticky gums harden on contact with the air to form a clear, yellowish substance known as **copal**. Sometimes trees are tapped to get this **resin**, but the best copal has hardened over a long period in the ground and is semi-fossilized. It is used mainly in paints and varnishes, but some kinds have a medical value, especially in treating skin diseases. The collection of copal and similar gums in the forests of South America is by people working for themselves, who may travel long distances between the trees that they wish to tap. It is unlikely that it will ever be a well-organized industry, because plastics have largely taken over from the natural gums.

Turpentine

Pines of cooler countries are tapped for turpentine, which is used as a solvent and for making paints and polishes. The resin that is left after distilling the turpentine is used for making many things, from floor covering to greases and printers' inks.

Mastic

Some plants produce a sap that does not dry completely but remains rubbery. The mastic produced by the lentisk pistachio tree, for example, is used for making temporary fillings for teeth.

Rubber

Gum trees first became important when people were looking for possible sources of rubber. Wild rubber was discovered in

▲ *Amber is a fossilized resin that has been formed from the gum exuded from conifers. Insects are often found embedded in it.*

The resin of pine trees is collected in cups. Oil of turpentine is made from the resin by distilling it either directly or with steam.

Brazil in the early 19th century and was used at first for waterproofing. When the process of **vulcanizing** was discovered, the demand for rubber increased and the **latex** of all sorts of plants was used. In 1895 the seeds of rubber trees were smuggled from Brazil to England, where they were germinated. These seedlings were

MAKING RUBBER FROM RAW LATEX

1. Tapping latex from a rubber tree.

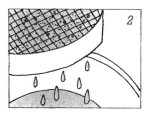

2. Straining latex through a fine sieve to remove dirt.

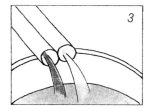

3. Latex poured into vats and coagulated by mixing it with acetic or formic aid. Water squeezed out of the coagulated latex.

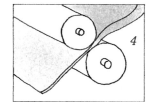

4. Latex flattened into a thin rubber sheet. These are dried in a smokehouse.

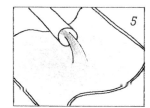

5. Raw rubber mixed with chemicals, mainly sulfur, to vulcanize or harden, it.

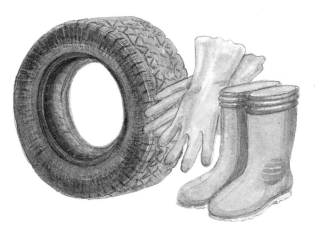

CHEWING GUM

Most chewing gums are made from the sapodilla tree, which comes from Central America. A series of long cuts are made in the bark and the gum is collected in a pouch at the base. In Guatemala most of the gum is taken from plantation trees, but elsewhere collectors tap wild trees in the forest. Several relatives of the breadfruit are also used on a small scale for making chewing gum (the same species are often used for making bird lime.)

shipped to Ceylon and sent from there to Malaysia and Indonesia. The Far East now produces 90 percent of the natural rubber crop. The trees are grown in plantations and are first tapped when they are about five years old, though they give their highest yields from about their twelfth year. A high yielding variety of rubber tree gives about 1 ton of latex in a year. In spite of this, natural rubbers have declined in importance since about 1945, when synthetic rubbers were first discovered and developed.

Cosmetics and Spices

Most people enjoy the scent of flowers and since early times dried leaves and flowers have been used, though their scent tends to fade quickly. The leaves of woodruff, for example, were put with linen to give it a fresh smell, and dried petals of roses and other flowers are the basis of potpourri used to scent rooms.

It is more complex to extract the essential oils that carry the scent of flowers or leaves. However, this is done on a large scale by the perfume industry, which grows roses, lilies, jasmine and other highly scented flowers.

Cosmetics

The cosmetic industry is now a very large one, and a great many vegetable oils are the basis of creams, both for cleansing and softening the skin. Palm oil, jojoba oil, olive oil and almond oil are among those used. Powders often include starches from maize and other cereals.

Spices

Highly scented plants often have a powerful flavor. We use such strong-tasting herbs and spices to give an extra zest to our food. Most spices come from the tropics. Some, such as cinnamon, which is the bark of a bushy tree, have been known and valued since Roman times. Later, the wish to get spices led to great journeys of exploration as people tried to find new routes to India and the Moluccas, which they called the Spice Islands.

Many parts of plants are used as spices. Cloves are unopened flower buds; nutmeg is a large oily seed and mace the membrane that surrounds it. Cardamom is a fruit. So are peppers, which are the berries of a woody climber. Vanilla is the seedpod of an orchid, and ginger is an underground stem or **rhizome**. Most of these plants are now grown in plantations. Many of them have to be harvested and prepared with great care so as not to

▲ Roses growing at Grasse in the south of France, center of the perfume industry. It takes 200 million rose petals to make about 32 fluid oz (1 liter) of rose oil.

▼ Lavender is also grown commercially and is used to prepare oils and perfumes.

SOAP

*T*hroughout the world people have found plants that they could use for washing themselves and their clothes. Soapwort is one of them. The lather made from the crushed and boiled leaves is quite effective. In the Middle Ages it was used in the dying process for woollen cloth, and in some places the sheep were washed with it before they were shorn! Soapwort was taken from Europe to America and can often be seen in clearings and growing along roadsides.

Other soap plants include the soap trees, whose berries make a lather. In some other species the pith of the bark, or the root, is used. In commercially made soaps and shampoos, oils distilled from many plants are used. These provide the smoothness and the fragrance that we expect to find.

damage the crop. Vanilla, which is one of the most difficult, is now replaced in many kitchens by synthetic vanillin. Although cheaper, it does not have the full flavor of true vanilla, which probably still has a good future.

▼ *Most spices are made from tropical plants.*

pepper

nutmeg

vanilla

clove

mace

Tropical Medicinal Plants

The fact that nowadays people live longer and healthier lives is partly due to our knowledge of tropical plants. Very many of them provide drugs that can reduce pain or prevent disease.

One of the most important is the *Cinchona* tree, which grows in upland tropical forests in South America. The bark produces several alkaloids, one of which is quinine. The trees are grown in small plantations and give their best drug production when they are over 10 years old. Another drug produced by the same trees is used to help people with weak hearts.

Rauwolfia

A drug more recently recognized as being useful comes from a large evergreen shrub called *Rauwolfia*. This grows in India, where the roots have been used since ancient time. Within the last 25 years it has been discovered by Western medicine and it is so valuable that a search for similar plants has led to the use of related species from Africa and America.

Yams

Yams have been grown for food for many years, though not all species are edible. The juice of some kinds has been used by local people to make arrow poison. However, scientific interest in them has increased with the discovery that they can provide the basis for sex hormones, birth control pills and anabolic steroids.

Poisons

The most famous of all tropical plant poisons is curare, which is still used in some places to poison the tips of arrows. Curare comes from strychnos trees, which as their name suggests, also produce strychnine.

Other powerful poisons, though not eaten, affect our health indirectly. The

▲ *The opium poppy has been used for centuries to make strong pain-killing drugs.*

best known of these are derris and pyrethrum, which are powerful insecticides. The value of these over some other insect killers is that they do not harm warm-blooded creatures. However, they have to be used with care for they are deadly to fish and amphibians, and are liable to kill useful insects along with the pests.

TEA AND COFFEE

*A*ll living things make waste products. In plants, some of these are known as alkaloids. Some alkaloids are liquid and are held in the sap, or they may be in the form of solid crystals. Often the plants make use of alkaloids as part of their defense against animals, for most of them are extremely bitter and some, such as strychnine and nicotine are highly poisonous. We enjoy drinking coffee and tea because both contain an alkaloid called caffeine, which acts as a stimulant on the muscles, heart and central nervous system. This is the reason why when we are tired, a cup of tea or coffee makes us feel brighter, though too much is harmful. Many alkaloids are so powerful that they are used in medicines.

▼*Many tropical plants are used to make drugs. The bark of the cinchona tree produces quinine, which is used to treat malaria and other fevers.*

The alkaloid reserpine is obtained from Rauwolfia. It is used in the treatment of mental illness and hypertension.

Cephaelis produces the alkaloid ipecacuanha, which is used to control amoebic dysentry.

Strychnine comes from the strychnos tree. Tiny doses of this deadly poison were at one time thought to be a good tonic and stimulant.

Natural Medicines

Plants from cooler parts of the world may also help humans by providing medicines. Many of them have names that tell us of the uses that people found for them.

Woundwort and self-heal were both used to cure cuts and injuries. Lungwort was given to people with coughs, and eyebright was thought to improve the sight. Milkwort and milk thistle were both prescribed for nursing mothers.

Ancient cures

American Indians used preparations of snakeroot to cure stomach pains and snakebite. They gave the name kinnikinnik to a special kind of dogwood, the bark of which had many uses, from making a tooth powder to a treatment for indigestion. In New Zealand the leaves of a tree called *Coprosoma* were used to treat cuts, sores and bruises, as well as many kinds of stomach illness.

Nature's remedies

In some cases the cures were imaginary – eyebright for instance was thought to look like the eyes of a person who was ill – but often the plants contained alkaloids that really do help sick people. One such plant is willow, for the bark contains salacin, the basic ingredient of aspirin. To this day, willow twigs are collected in some parts of the world to make local cures for fevers and rheumatism.

Sometimes the plants used were deadly poisonous and were often used to remove worms or head lice. One with a different function was the deadly nightshade. This was sometimes known as belladonna, which means "beautiful lady." In the past ladies in France and Italy used to put the juice of deadly nightshade into their eyes, to open the pupil and make their eyes look large and alluring. Nowadays, an ophthalmologist may use a drug made from this

▲ *Digitalis made from foxglove is used to treat heart disease.*

▼ *Drugs made from deadly nightshade are used to treat eye and heart disease and a nervous disorder called Parkinsonism.*

PENICILLIN

It is not only large flowering plants that are of medical use. One of the great discoveries of the last 50 years is the use of the mold penicillin as an antibiotic. Many diseases caused by bacteria, such as pneumonia, are no longer the scourge that they once were because of the use of antibiotics. In the laboratories in which such discoveries are made, other plants play a part. Some seaweeds are perhaps the most unexpected of these, but they are vital, for they provide the sterile gel on which cultures of experimental molds, bacteria or other small organisms are grown and tested.

Sir Alexander Fleming discovered penicillin in 1928

plant or from thorn apple. As it makes the pupil open wide, the eye can be examined more easily.

Today, the pharmaceutical industry is able to synthesize many drugs in the laboratory. By doing this, they can be sure of the strength and purity of the product. But in some cases the drugs still have to be obtained from plants. Deadly nightshade is one of the most valuable, not only for eye specialists, but because drugs used to treat heart diseases and Parkinson's disease can be made from it. Digitalis, extracted from foxglove is used as a heart stimulant. Arnica, a plant related to the dandelion, is also useful. An alkaloid made from the roots is especially valuable for treating bruises and strains. Witch hazel can be used to help bruising and soothe tired eyes.

oxeye daisy leaves are used to treat catarrh

viper's bugloss has been used to treat fevers

common comfrey leaves can be used on sprains

coltsfoot is used to treat coughs

Plants for Pleasure

Human beings have probably always enjoyed plants for the beauty of their color, shape and scent, as well as for their practical uses. From the time that people began to live in towns and villages, they planted gardens.

At first, the flowers grown would have been taken from the wild country nearby. The biggest plants with the brightest colors were probably chosen by the gardeners, and in time cultivated plants changed as a result of this selection. Today, most garden plants are larger than their wild relatives.

We still grow many of the same plants as our ancient ancestors. Violets, roses and lilies, for example, would be familiar to a Roman or mediaeval gardener. But if it were possible for such a person to see the gardens of today he would be surprised at the variety. In 1629 a gardener wrote of the great variety of roses that he grew. He grew only 30 varieties. Today, we have hundreds of different varieties.

New discoveries

There are many plants that would be completely unknown to early gardeners. Some, such as dahlias, named after the Swedish botanist Dahl, and fuschias, named after Leonard Fuchs, a German professor, come mainly from Central and South America. Other plants were first found in places explored much more recently. Towards the end of the last

◄ Two plants that you can find in many gardens are Lewisia (right) and Clarkia (left). These are named after Meriwether Lewis and William Clarke, who were two of the first explorers of the American West, where these flowers are found growing in the wild.

century and in the early 1900s, collectors went to the Himalayas and the mountains of the Far East. They brought back a galaxy of plants, including new species of peonies, primroses, rhododendrons and many others.

Tropical plants

More recently still, a vogue for tropical plants has developed. These need warmth throughout the year and cannot stand frost, so they must either live in a heated greenhouse, or be kept indoors. The Swiss cheese plant is a common houseplant for it will stand being cut back, though if it is allowed to grow unhindered it will eventually produce delicious fruit. Smaller plants, often with insignificant flowers but attractive leaves, are most popular. These include mother-in-law's tongue, from West Africa, and the zebra plant from Brazil.

Garden centers

A major industry has developed in recent years to cater for our need to beautify our surroundings with plants. Nurseries and garden centers use modern scientific techniques to grow huge numbers of plants of named varieties that are exactly the same as each other. The control of weeds and pests by chemical means has made growing plants simpler than ever before. Perhaps as a result of this, and because many wild plants and animals are endangered, people have been returning to the small wild species once more. It is now possible to have a garden of modern plants and a wild garden, side by side.

◄ Many ancient societies made elaborate gardens with trees for shade and flowers for color. This is still the basic idea of most gardens.

▼ Nurseries and garden centers grow many unusual kinds of plants.

Plants and the Future

Every day the human population of the world gets larger. People need more food and more places to live. People depend on plants for most of their food, so scientists are working to find new crops or better ways of growing old ones.

Scientific breeding programs have increased the yields of crops such as wheat and rice hugely during the last 50 years. It is possible that this improvement will continue into the future.

New crops

New crop plants include a kind of grass called *Echinochloa*. It is used mainly as animal fodder in Africa but in the Far East people sometimes eat the grains, for it will thrive where rice cannot. In Australia, another kind of *Echinochloa* grows in very dry, arid areas. This grass may be used in the future as a valuable food crop for the deserts of the world.

Many of the possible improvements to life in the future depend on the development of new crops from wild species. There is, for example, a kind of wild coffee

▶ A few flowering plants grow at the edge of the sea. Some, such as eel grass or rice grass, may become crops of the future, for both produce usable grains. Duckweed, which grows on stagnant water, could be harvested to feed farm stock, and some microscopic plants may be used as human food in the future. These have the advantage that they can be grown in factories.

crambe

▲ Seaweeds can give us minerals and gels, but they are not suitable as a staple food.

kelp

duckweed

that does not contain caffeine. If this could be made more productive, or crossed with existing kinds of coffee, it might be possible to have a natural decaffeinated drink, which would be better than the present types of coffee.

Genetic engineering

Some new plants may be laboratory made, by **genetic engineering**. We hear a lot about this, but the steps needed to make useful new plants are very small and slow. At the moment, for instance, an experiment to improve tomatoes by changing one gene may result in fruit that keeps longer and better. Another experimental dream is to combine grain crops and peas, but a major breakthrough would be needed for this.

New drugs

There is a constant search for new drug crops to improve the health of humans and their animals. Many old remedies using herbs and plants are being investigated. A chinese cucumber is a possible source of drugs for treating AIDS. Another eastern plant, a kind of forsythia often grown in gardens, is used for many medical purposes in China and Japan, including the treatment of breast cancer. A cure for either of these illnesses would be a great step forward in world health.

THE GREEN REVOLUTION

In recent years many forests and wild places have been destroyed so that cattle can be kept there, or so that traditional crops, often not well-suited to the areas, can be grown. Often we do not know what is being destroyed, and many plants are near to extinction unless we can call a halt to the destruction. Fortunately, many people have at last woken up to the danger. Green – the color of plants – is the name given to ideas that are now becoming popular in politics as well as science. The basis of these ideas is that we must live in partnership and harmony with our surroundings. If we can do this, then our future, with many practical plants, will be more secure.

▼Spirulina, *a blue-green alga, is now cultivated in Mexico. It is dried and the protein is added to various food products.*

Glossary

Agricultural revolution The change in people's lifestyle from gathering wild plants and hunting wild animals to living in settlements, growing crops and keeping animals. This first happened about 10,000 years ago.

Algenates Jelly-like substances obtained from seaweeds.

Awn A bristle on the husk that protects the grains of some grasses and cereals.

Breed true A plant that breeds true is one whose seeds produce plants like the parents in all respects.

Bulb A special shoot used by some plants to store food for next year's growth. It consists of a short stem surrounded by fleshy leaves

Chipboard A wood substitute, made of chips of wood, bonded together with synthetic resin.

Cob The hard central part of a head of maize, on which the seeds grow.

Copal Hardened resin produced by various trees, used in making varnish.

Crown The crest of leaves on the top of a pineapple.

Ergot A disease of rye, which can cause death to humans and livestock.

Fern A large leafy plant which does not have flowers but reproduces by means of spores.

Funfural A solvent extracted from the husks of oats.

Fungus (*plural fungi*) A non-flowering plant that does not make its own food, but is able to use nutrients from other plants or from animals. The fruiting bodies of large species are known as mushrooms and toadstools. Small ones include mildews, molds and rusts.

Genetic engineering Altering the chemical pattern of the genes, which are the blueprint of a plant or animal, held in the nucleus of its cells. This results in new forms of living things.

Gluten A sticky substance in flour.

Grain The seed of a grass or cereal. The word is sometimes used to mean the crop.

Grafting A method of growing plants that do not breed true by attaching a growing shoot to the rootstock of another, related plant.

Hardwood The wood of broad-leaved non-coniferous trees.

Horsetail A non-flowering, spore producing plant, with whorls of tough, narrow leaves.

Husks Small leaf-like organs that protect the seeds of grasses and cereals. As the seeds or grains ripen, the husks become dry and hard.

Hybrid A cross between two species. It is often large and strong, but sterile and unable to breed itself.

Kernel The seed of a nut or a fruit with a stone, like a plum. Sometimes used for grain seeds.

Latex A rubbery substance produced by some plants.

Lichen A non-flowering plant made up of cells of a fungus and an alga living together.

Minerals Non-living materials many of which are needed for growth. They are obtained by plants from the soil.

Mosses Small non-flowering plants that usually live in damp places. They reproduce by means of spores, which grow in pepper-pot like capsules.

Mutant A plant or animal in which the chemistry of the genes, which control what the organism will be like, has been changed, either by accident (a mutation) or deliberately (genetic engineering).

Nitrogen A gas that forms about 78% of the air. We do not use it in breathing, but it is very important to all living things, for it forms compounds that are part of proteins and other living matter.

Pigments Coloring matter.

Proteins Very complicated molecules found in all living things and vital for many functions of life. Plants are able to make their proteins from non-living minerals; animals cannot do so.

Pulses The seeds of peas, beans and related plants.

Rachis The central stem of a flower head or a leaf with several leaflets.

Resin A waterproof, sticky substance produced by many plants. It is often exuded from a cut in the stem.

Retting Soaking plants such as flax in water, to get rid of the soft tissues.

Rootstock The roots and lower part of a strong growing plant on which another more desirable plant is grafted.

Sap The fluid, containing minerals and sugars, that circulates in plants.

Scurvy A disease caused by lack of vitamin C in the diet.

Shingles Thin pieces of wood used like tiles on buildings.

Softwood Wood obtained from coniferous trees.

Starch A compound formed of carbon, hydrogen and oxygen. It is used by plants mainly for food storage. Many of the plants that we eat, such as potatoes, contain a lot of starch.

Suckers Shoots that arise from the roots or underground stems of plants.

Tuber An underground stem or root, swollen with stored food.

Veneer A thin layer of decorative wood used to beautify furniture made of poorer wood.

Vitamins Organic substances needed by animals in very small quantities to ensure normal development and growth. Lack of vitamins, which cannot be synthesized by the body, leads to deficiency diseases.

Vulcanizing A method of treating rubber with sulfur at a high temperature. This makes the rubber much stronger and more elastic.

Index

Further Reading

Young Adult Books – Practical Plants

Dineen, Jacqueline. *Cereals*. Hillsdale, New Jersey: Enslow Publishers, 1988.

Dineen Jacqueline. *Sugar*. Hillsdale, New Jersey: Enslow Publishers, 1988.

Dineen Jacqueline. *Cotton & Silk*. Hillsdale, New Jersey: Enslow Publishers, 1988.

Johnson, Sylvia. *Rice*. Minneapolis: Lerner, 1985.

Young Adult Books – General

Black, David. *Plants*. New York: Facts On File, 1986.

Forsthoeful, John. *Discovering Botany* New York: DOK Publishers, 1982.

Lambert, David. *Vegetation*. New York: Franklin Watts, 1984.

Adult Books – General Reference about Botany

New England Wild Flower Society Staff. *Botany for All Ages*. New Jersey: Globe Pequot Press, 1989.

Rost, Thomas L. Botany: *A Brief Introduction to Plant Biology*. New York: John Willey & Sons, 1984.

Tootill, Elizabeth (ed.). *The Facts On File Dictionary of Botany*. New York: Facts On File, 1984.

Adult Books – Practical Plants

Stoskopf, Neil. *Cereal Grain Crops*. New York: Prentice Hall, 1985.

Photographic credits

t = top, *b* = bottom, *l* = left, *r* = right

Cover: Frank Lane/Holt Studios; page 6 Bruce Coleman/Hans Reinhard; page 8*t* Frank Lane/G. Dodd; page 8*b* Bruce Coleman/Eric Crichton; page 10*l* Frank Lane/Holt Studios; page 10*r* Anne Ronan Picture Library and E.P. Goldschmidt and Co Ltd; page 12*t* Frank Lane/Holt Studios; page 12*b* G.S.F. Picture Library; page 13 C and L Nature World/Cyril Laubscher; page 14 Frank Lane/Holt Studios; page 15*t* Frank Lane/Holt Studios; page 15*b* Frank Lane/Holt Studios; page 16*t* Bruce Coleman/Eric Crichton; page 16*b* Bruce Coleman/Eric Crichton; page 18 Bruce Coleman/Norman Myers; page 19 Frank Lane/Holt Studios; page 20 Bruce Coleman/Eric Crichton; page 21 Frank Lane/Holt Studios; page 22 Bruce Coleman/Eric Crichton; page 23 Frank Lane/Holt Studios; page 25*t* Frank Lane/Holt Studios; page 25*b* Frank Lane/Holt Studios; page 26 Frank Lane/Holt Studios; page 27 Frank Lane/K.G. Preston-Mafham; page 28*t* Frank Lane/Holt Studios; page 28*b* Bruce Coleman/L.C. Marigo; page 31*t* Bruce Coleman/A.J. Deane; page 31*b* Bruce Coleman/Hans Reinhard; page 32 Tate and Lyle; page 33 Bruce Coleman; page 34*t* Bruce Coleman/Norman Myers; page 34*b* G.S.F. Picture Library; page 35 Smith/Polunin Collection; page 36*t* Bruce Coleman/M.P. Price; page 36*b* Bruce Coleman/Roger Wilmshurst; page 37 J. Pope; page 38*t* Michael Holford; page 38*b* Bruce Coleman/Gerald Cubitt; page 39 Bruce Coleman/Jane Burton; page 40 Robert Harding Picture Library; page 42*t* Frank Lane/Holt Studios; page 42*b* Frank Lane/Holt Studios; page 45*t* Bruce Coleman/M.P.L. Fogden; page 45*b* Bruce Coleman/Charlie Ott; page 46 Bruce Coleman/G. Ziesler; page 47 Smith/Polunin Collection; page 48*t* Frank Lane/Holt Studios; page 48*b* Bruce Coleman/H. Kranawetter; page 49 Bruce Coleman/Prato; page 50 Bruce Coleman/Eric Crichton; page 51 The Tea Council Ltd; page 52*t* Bruce Coleman/Frans Lanting; page 52*b* Bruce Coleman/Hans Reinhard; page 53 Mary Evans Picture Library; page 54 Bruce Coleman/Eric Crichton; page 55 Bruce Coleman; page 56 Frank Lane/M.J. Thomas; page 57*l* Dr Hall, Kings College, London University; page 57*r* Bruce Coleman/Alain Compost.

M